Don't Read This Unless You Want To Pass Your Math Class: a guide to success in college mathematics courses

By Angie Schirck-Matthews

Table of Contents:
Introduction
    Chapter 1: Learning Styles
    Chapter 2: Attitude Adjustment
    Chapter 3: Practice Makes Permanent
    Chapter 4: Note Taking
    Chapter 5: Study Tips
    Chapter 6: Test Anxiety
Appendix A

About the author: Angie Schirck-Matthews is a Senior Professor of mathematics at Broward College where she has been teaching college level mathematics since 1991. She is also co-author of the second edition of Math in Our World, A Liberal Arts Math textbook. She received her undergraduate degree in mathematics from Florida Atlantic University and her graduate degree at the University of Miami. She is also seeking a second graduate degree in Geospatial analysis at the University of Florida.

Introduction

I see you were willing to cough up a few George Washington's to read this book; I guess that means you want to pass your math class. Good, that's a great start! But wanting and a penny will get you an F. I'm going to cut to the chase here, I don't have a magic potion that, if you drink it, you are guaranteed to pass; the truth is, if you want to pass your math class, you're going to have to do some work! So if you're not willing to put in a little effort, quick, return this book now! But if you are willing to work for it, what I have to say can help you succeed. Success in mathematics is up to you. Your teacher's job is to convey concepts in a way that is understandable to you, your job is to study, practice, and learn these concepts. It doesn't matter if you have a "good" teacher or "bad" teacher; the teacher does not determine your success or failure, you do. Even if you have a "bad" teacher, you will pass (notice I said will, not might), if you are willing to put in the work and follow my suggestions. "Good" and "bad" are in quotes above because these words are subjective. I read end-of-course reviews from students (after grades are turned in of course) and sometimes I wonder how one student was in the same class with all the others. The dichotomy of their responses is amazing. The point is, what one student considers a "bad" teacher may be "good" for someone else, since we all have different learning styles.

Chapter 1: Learning Styles

There are three basic learning styles: Auditory, visual, and tactile. Determining your learning style will help you, not only in your math class, but in all your classes. A quick Google search for learning style quizzes will get you a plethora of results. Take a learning style quiz to determine how you best learn. When you know your learning style read the following tips for your particular learning style. If you find you have a mix of learning styles, follow the tips for all that apply.

Auditory Learners. Since you learn better by hearing, you should sit near the front of the class so you can hear the teacher. If you're having trouble hearing, have your hearing checked! Keep your ears clean and free of wax. Ask your teacher for permission to record lectures (she may say no and you must respect this). Loud noises will distract you, so find a quiet place to study. Do not study while listening to music or with the TV on, as this will distract you as well. Also, when you study, read all study materials, assignments, and directions *out loud*. Use flash cards and put key terms on them, then read them out loud, or ask a friend to read them to you. You might also want to record yourself reading your notes and listen to the recording. During tests quietly read the questions out loud to yourself.

Visual learners. Since you learn better by seeing, you should sit near the front of the class so you can see the teacher and the board. If you need glasses use them, and make sure your prescription is up to date. Use online videos to help understand topics that you need more clarification on or that you didn't get during the class lecture. You will be distracted by visual things, so don't sit near windows in class or while studying. Study in a well-lit, distraction-free environment; make sure the TV is off! Put key words or terms on flashcards and look through them repeatedly. Take written notes in class and invest in a good set of colored pens or pencils to color-code things; read through the notes repeatedly. This learning style is my personal learning style and as a student, when I repeatedly read through my notes, I could "visualize" the explanations written in my notes in my mind during the exam; this helped me to recall needed information. Draw pictures to help explain or understand concepts.

Tactile Learners. You learn by touch (but touching the teacher is not advised), fidgeting is your thing; while you study, fidget away! Shake your foot, tap your pencil, click your pen, chew gum, rock in a chair, walk around, stand on your head if you want, whatever works to keep your mind on your task, do it. Write down important concepts and formulas on flash cards and arrange them to show relationships between concepts. Trace out key words or formulas repeatedly with your fingers. Look for hands-on manipulatives to help understand concepts. My daughter's third grade teacher had a scale that the kids put blocks on and took them off to understand balancing algebraic equations. Ask your teacher if he has any manipulatives such as this or look for some in an education supply store. Take written notes, the act of writing will engage your sense of touch. Use a computer while studying to activate your sense of touch, this will help reinforce learning. You learn best by doing, as opposed to other learners who learn by seeing or hearing.

Chapter 2: Attitude Adjustment

That's right, I said you need an attitude adjustment! In all my years of teaching, if I had a quarter for every time I heard a student utter the words "I hate math," "I suck at math," or "I've never been good at math," I'd retire quite wealthy! Maybe I *should* start charging; I'll make an attitude-check jar, like a swear jar. But I digress; back to the topic at hand.

My dad is a pilot of small aircraft; he used to take me flying as a kid. He would explain to me what all the instruments on the dash were and how they were important to the safety of those in the airplane. Every aircraft is equipped with an attitude indicator (pictured).

This little instrument is vitally important because its job is to indicate the orientation of the airplane relative to the horizon (i.e. the earth). If the attitude of the plane is not adjusted properly, the plane will not go in the right direction, and could crash. If the plane's attitude is not right, it will fail! As with the airplane, your attitude is crucial to your success in math; without the right attitude you could possibly "crash," mathematically speaking. It's hard to overcome a lifetime of believing that you "can't do math," but if you want to succeed, overcome you must. In order to do this you need to tell yourself you are good at math, you can do math, and never utter the negative affirmations that you've believed your whole life. Motivational speaker Zig Ziglar said, "It is your **attitude**, more than your **aptitude**, that will determine your **altitude**." How high do you want to fly? Change your attitude and you can fly high (and in the right direction).   This is going to sound corny, but I want you to speak the following positive affirmations out loud every day until you believe them. Write them down and tape them on your bathroom mirror so that you can see them every morning.

I can do math.

I am good at math.

In fact, I am a math genius.

Math is easy.

Math is fun.

I love math.

I need math.

My math skills are improving every day.

I'm so good at math, I might become a math teacher.

    Make your own attitude-check jar, and every time you make a negative comment about your mathematical ability, put a quarter in the jar. Each class period give the money from your attitude-check jar to your math teacher. In short, keep an eye on your attitude indicator so that your metaphorical plane is going in the right direction.

Chapter 3: Practice Makes Permanent

You have probably heard the old saying that "practice makes perfect." Unfortunately, that saying you've heard all your life is not accurate! Only *perfect* practice makes perfect. Have you ever driven to school or work (or anywhere you go often) and find yourself on a road along the way and say to yourself "I don't remember driving here"? That's because you have driven that path so many times that the way is made permanent in your mind. Perhaps you were singing along to the radio, or lost deep in thought, but your subconscious knew where you were going and how to get there. But suppose for some reason you repeatedly attempted to drive to school, but you were doing it all wrong. In fact, every time you set out to school, you ended up at the mall instead. So the next time you're lost in thought along the road on your way to school, where do you suppose you're going to end up? Exactly, the mall, because all that "wrong practice" was made permanent in your subconscious. I play on a softball team, and I watch women during practice catch a ball in the outfield and then stand there holding the ball and ask "where was the play?" I repeatedly remind them, "know where you're going to throw the ball before you catch it, because what you do in practice is what you're going to do in the game." You can't catch the ball and just stand there while the runners run around the bases! The same is true with learning mathematics. Practice makes permanent; only perfect practice makes perfect! What you do in practice is what you're going to do on the test!

Now that we have that all sorted out, exactly how do you go about practicing math perfectly? Good question! The first step is to study the concepts *before* you attempt the homework. It's not sufficient to just watch the teacher lecture, and then go and attempt the homework. I've read from multiple sources that we only remember about 20% of what we hear, and 30% of what we see. This means if you listen to the lecture and watch your teacher, you're only going to remember about 30% of what was done, at best. If you actually write (take notes, which is highly recommended, see note-taking discussion later on) this percentage more than doubles. But it's still not enough for "perfect practice." So, before you attempt the homework, review your written notes. Also, read the text. Yes, I said READ THE TEXT! The text is full of explanations and examples that might be different from what your teacher did in class; the author may also explain something in a way that is more understandable to you. Do this especially if you're a visual learner. If you're an auditory learner, read the text aloud. If you're a tactile learner, get a real copy of the text that you can hold in your hands while you read. If after reviewing the notes and reading the text you're still unclear on some concepts, get help! Take advantage of your instructor's office hours. If your school has a tutoring center, use it! Use online videos, often textbook publishers provide these free of charge with your text book or e-book purchase. If not, there are a variety of videos online at websites like [Khan Academy](#) and [YouTube](#). The bottom line is, make sure you *clearly* understand the concepts *before* attempting the homework.

    Alright, so now that you're sure you understand the concepts, it's time to practice! But remember, we want to practice perfectly, because what we do in practice, is what we will do on the test. This means if you're working a problem and you get stuck or you're not sure what to do next, STOP! Do not continue, do not guess at the next step; get help. This is why it's a good idea to do your homework in a tutoring center if your school offers one.

Where I teach, students can borrow laptops from the Academic Success Center and take them to the Math Lab or STEM center. If your using online homework and don't have your own laptop, see if this is offered at your school. If you don't have access to a tutoring center, use the resources mentioned previously. Most online homework systems have a help button as well, so use that if you need it. But beware, excessive use of online help buttons isn't going to help! I once had a student that got a perfect score on every homework (and she *was* doing her own work) but failed every test! I asked her about her study habits and I discovered she got through the online homework by hitting help at every step and mimicking what she saw, the problem is she was getting through the homework this way, but she wasn't learning! Her practice included the excessive use of help buttons so she "needed" them to succeed. Unfortunately, when she got to the test she didn't know what to do because her "help button" wasn't there. So, help buttons are great if you get stuck, but try to understand the concepts first and don't use them as a crutch. So go, and practice, practice, practice, but remember only PERFECT PRACTICE MAKES PERFECT.

Chapter 4: Note Taking

Taking written notes in class is important for all learning styles. How you take notes is important too. In high school I had an algebra teacher who used an overhead projector (for those that don't know what that is, it's like PowerPoint before computers). She would have everything prewritten on the transparencies and explain though it quickly. We students would be rushing to write every word and example on the transparencies before she switched to the next one. As a result, we were not paying much attention to the teacher or her explanation; we were too busy trying to write everything down. There needs to be a balance in your note taking. You don't need to write everything down. Let me repeat, you do not need to write everything you see on the screen or board down! Determine what's important and write it down, but also listen to and watch what the teacher is doing.

In a math class, it's not necessary to write down definitions, theorems, or steps in problem-solving processes. These things can be easily found in the text book; just make a notation in your notes to look them up later. When I put definitions and such on the screen, I tell my students "don't write these down, they are in the book and on the slides and you can look them up later, listen to me instead." I also provide the PowerPoints or notes that I use in class for them in our learning management system so they can bring them to class and take notes on them. Some students do this using their tablets and others print them out beforehand. If your instructor offers such a thing, take advantage of it! If your teacher allows it, use your cell phone to take a picture of the definition rather than writing it down. The important thing is to listen to the teacher and jot down notes on the explanations the teacher gives that are not on the slide; this will help you understand the meaning.

After your instructor explains a concept she will most likely give you some examples. Write down these examples while listening to your teacher. Be sure to write everything *exactly* as your teacher does (even if you don't understand it, which, by the way, if you don't understand something it's okay to stop and ask your teacher to clarify), this will get you in the habit of writing mathematics properly. The way a teacher writes the problem on the board is the way he's going to expect you to do it on the test. Proper notation in mathematics is crucial and if you write it in your notes improperly, well, we'll have that imperfect practice thing going on again. Jot down explanations as the instructor gives them. See the note-taking example in **Figure 1** in Appendix A.

Notice that the steps for solving a linear equation are on the left, with notes to the right. The notes are what the teacher said while she was doing the example.

Write down important formulas and things you need to remember. Try to pay close attention to important things the teacher says, especially things she repeats, and write these down. I usually stress important things to my students by saying things like "you will see this again," or "this is important," or "you need to know this." Listen for clues like this to important concepts of which you need to take note.

Keep *three* notebooks for your math class. One for in-class notes, one for after-class notes, and one for practice homework. Also, when scheduling your classes try not to schedule a class immediately following your math class. This way, immediately after class, you can take out your after-class-notes notebook and rewrite your notes while what you did in class is still in your short-term memory. As you're rewriting your notes add in the definitions, theorems, and problem solving steps you did not write down in class. Again, these should be easy to find in your textbook, or if you took a photo they are readily available for you. The rewriting of notes serves multiple purposes. First, and most importantly, it's reinforcing what you learned in class; second, it's giving you a chance to organize your notes and write them neatly; and third, it's giving you a well-written, organized study guide. This is also a good time to make flash cards of the important formulas or definitions you learned in class and will need to remember. Figure 2a in Appendix A shows an example of in-class notes. There's a reminder at top to look up the definition of a linear equation, followed by a comment the teacher made, then a reminder to look up the steps for solving a linear equation, and finally the teacher's worked out example with notes to the side. Figure 2b in Appendix A shows an example of the after-class notes. Here the full definition of a linear equation is written out, followed by the teacher's comment, then the steps for solving a linear equation written out fully, and finally the example the teacher did will be written out completely with notes to the side.

In your homework notebook, even if you're using an online homework system, do your homework problems in an orderly fashion, as you instructor did in class, writing out steps clearly and neatly. Do not use scrap paper to do your online homework, this will get you in the habit of writing things incorrectly and you will also do so on the test. I tell my students if I can't read or follow their work, it will be marked wrong. I, and most instructors, do not want to look through messy work to see if we can find something right, so practice working neatly on homework because practice makes permanent!

Chapter 5: Study Tips

I've already mentioned quite a few study tips, but let me summarize them here and give you some additional suggestions. Learning math, or any new subject, can seem overwhelming. There are all these formulas and procedures, and your instructor assumes by next class you know them, because they are needed for the next lesson. Math is cumulative, that is, each lesson you learn uses what you learned in the previous lessons. You can't forget anything! When I teach Calculus 1 my students are usually very surprised by the amount of algebra and trigonometry they're expected to know. Usually the results on the first test are abysmal, and it's not because these students didn't understand the new calculus concepts I taught them, but because they've forgotten the pre-requisite algebra they need to know in order to complete the problems. The point is, you need to get what you learned in class out of your short-term memory, and into your long-term memory.

If you feel like you study hard but still don't do well, then you're probably not studying correctly. One study method that will certainly set you up for failure is "cramming for the exam." It may work *sometimes* to get you a passing grade on the exam, but you haven't learned anything and you *will* be expected to know it for the next chapter. The following study guidelines should help you learn and retain what you've learned.

Don't try to do too much at once. Shorter study sessions that really concentrate on one or two topics are more effective than long study sessions and trying to "learn it all." Studying for long periods can become exhausting and your mind will tend to wander and you will not be absorbing the material. I suggest you set a dedicated study time *each day* (a break between classes is perfect) even if it's only 20 minutes. Find a quiet, distraction free place to study, and if you start feeling exhausted or overwhelmed, it's time to stop!

Make a study plan. Set goals for each of your study sessions. Plan your study time each week by looking ahead at the topics you will be going over in class. See example study plan in Appendix A. Of course your study plan may vary based on due dates of assignments, but the idea is to have a plan, and try to stick to it. Once you figure out how your class works and what your teacher expects, making a study plan should be easier. You should study a little bit every day, this makes remembering what you learned easier. If you work well with others, plan study sessions with your class mates. The email tool in your school's learning management system can be used to send an email to your classmates to ask if anyone is interested in forming a study group. One of the best ways to learn something is to explain it to someone else. When I do in-class projects with my students I will often tell a student to whom I just explained something to explain it to the person next to them. This helps both students, but especially the one doing the explaining.

Study every day. I'm repeating what I said above because it's important. *Repetition* is important, especially when learning mathematics. Learning mathematics is like learning to ride a bike. When you first tried to ride a bike (without training wheels) you couldn't balance it, your mom or dad had to hold you up; but you tried over and over again and eventually you got it! After that, there was no stopping you, you rode your bike every day. And you know what's really amazing? You will NEVER forget how to ride a bike. You can go 10 years and not ride a bike, but the moment you get up on one, it will all come back. That's because practice makes permanent! The same is true with mathematics. If you study a little bit every day, eventually you'll realize that you actually KNOW it. It has moved into your long term memory, where it needs to be for you to be successful.

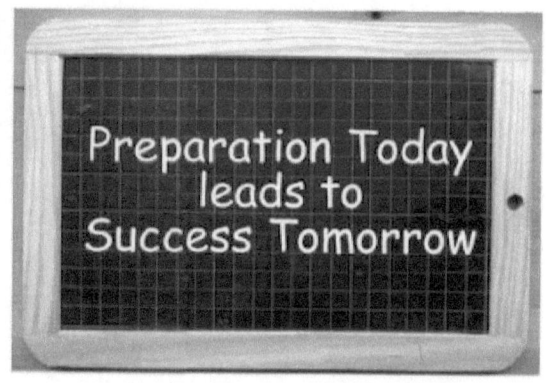

Prepare for class ahead of time. As I said above, repetition in mathematics (or anything you're trying to learn or remember) is important. If you want to improve your understanding while in class you should prepare for class ahead of time by reading the section in the text that you will be going over in class. You could also look up and watch videos on the topic, either in your textbook's online homework system, if available, or elsewhere online. If you prepare ahead of time, you will be able to ask your instructor appropriate questions to clarify gaps in your understanding.

Don't just study for the exam. Studying only for the exam will not put the topics in your long term memory. Test reviews are great if your teacher provides them, but they should not be the only thing you use to study. A test review should be used to practice as if you were taking the exam. Study daily as mentioned before, read through your notes daily, practice problems daily, then, after you've studied and are confident with the material, do the practice test, *as if it were a real test*, without looking at the answers. When your finish, grade yourself. This will be a good indication of how you will do on the exam. If you are not satisfied with your "grade" you can go back and study some more. If your teacher does not provide you with a practice test, most textbooks have a chapter test or chapter review that you can use as your practice test.

Study to understand. There's a difference between memorizing and understanding. You can get through an algebra problem by memorizing a set of steps, but if you don't understand what you're doing, when faced with a problem that looks different, you won't know what to do. When you're first working on a problem, follow the steps and write down WHY you're allowed to do that. If you're not sure you know why, look at the examples in the text, there's usually an explanation beside each step of the problem. You can also ask you teacher or tutor to explain WHY a particular step is allowed. For example, when you "build up" a fraction, you can multiply numerator and denominator by the same value as in $\frac{3}{4} = \frac{3 \cdot 4}{4 \cdot 4} = \frac{12}{16}$. WHY? Because $\frac{4}{4} = 1$ and multiplying by one doesn't change the value of the fraction. There are a lot of common mistakes I see students do repeatedly, like, for example, trying to solve an expression. You can't solve an expression, you can only solve an equation. If you *understand* the difference between an expression and an equation you will not make this mistake.

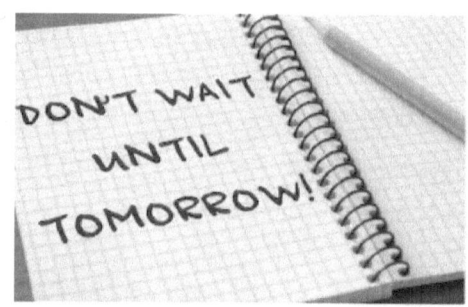

Do not procrastinate. If you wait too long to start studying you will find yourself "cramming for the exam." This is why I suggested that you not schedule a class after your math class. This is a perfect time to organize your notes and reinforce what you've learned. You should make an effort to learn the material taught in one class *before* the next class meeting. Remember that math is cumulative, and your instructor will expect you to know what was taught in the previous class in order to be able to understand what is being taught in the current class.

Make flash cards of key terms and formulas and look at them frequently, even when it's not "official" study time. You can keep them with you and look at them when you're on a break at work, or when you're sitting in another class waiting for the instructor to come in, or even during commercial breaks when watching TV. This is a good way to start memorizing formulas your instructor will expect you to know.

Chapter 6: Test Anxiety

I know many students suffer from test anxiety. It's normal to feel a little anxious during a test. When I was a student I had a personal rule regarding exams; I would never look at my notes or study the DAY OF the exam. If I didn't know the material by then, I knew it was too late. If you've studied correctly, using all the advice I've given up to this point, you should know the material. **Confidence** in your ability will help you overcome test anxiety. The day before the exam do your practice exam as I instructed previously (see the "Don't just study for the exam" section under Study Tips). If you did well on your practice exam, you will, *will*, WILL, **WILL** do well on the real exam (no, I'm not stuttering, I'm stressing), so be confident and relax!

Make sure you get a good night's sleep, at least 8 hours, the night before an exam. Humans cannot live without sleep. Our bodies need it in order to repair and recover from the day's activities. Not only will sleep make you feel better in the morning, but it will give your brain a chance to fuse memories of what you've studied and learned the day before. If you have trouble falling asleep, lay off the stimulants, such as coffee, tea, soda, or energy drinks. Also, do not drink alcohol the night before an exam! Not only might you be hungover the next day, but even a small amount of alcohol can interfere with sleep. Wear an eye mask if lights from outside or electronics inside your room bother you. Try using a white noise machine or listening to soft music while you fall asleep. Exercise in the evening can also help you to feel tired enough to fall asleep at bedtime. So, take an evening walk or jog, or ride your bike, or go workout, but not too close to bed time. Make sure there's at least a couple of hours between your exercise activity and bed time.

In the morning of the exam, eat a good, well-balanced breakfast. Just as sleep helps your body repair from the day's activities, breakfast helps your body prepare for the day's activities; and, believe it or not, breakfast can help improve your memory! Experts agree that breakfast is the most important meal of the day, so don't skip it, especially on a test day.

Okay, so you have studied, done well on your practice test, gotten a good night's sleep, and ate a good breakfast. You're sitting in the classroom waiting for the instructor to come in with the exam and you're still feeling a little anxious. Take a deep breath, in through your nose, out through your mouth. Do this several times and feel the calming effect. Close your eyes and visualize a calming scene, something that makes you feel peaceful, a boat on a lake, a mountain stream, flying in the clouds, etc. Give yourself some positive affirmations. I know the material, I've studied, I am confident in my ability, I will do well on this exam. If at any time during the exam you're starting to feel overwhelmed or anxious, stop, put your pencil down and take a minute to calm yourself again. Remaining calm and focused will help you succeed. And remember **YOU KNOW THIS STUFF, YOU'VE GOT THIS!**

In conclusion, I am not going to wish you good luck in your math class, because it's not about luck at all. It's about hard work, and effort! But I guarantee if you put in the effort and follow my suggestions, you WILL pass.

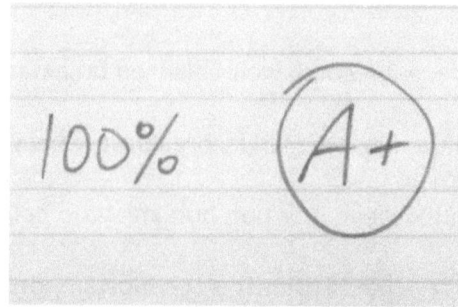

Appendix A

Sample Study Plan.

Monday – Class day. After class: rewrite and organize notes, make flash cards. Evening: re-read section 2.3 in the text; Goal: understand examples in text. *note to self: always carry flash cards with me and read through them at convenient times*

Tuesday – Study in the Math Lab. Go over section 2.3 notes. Goal: understand concepts in 2.3. Begin practice homework for section 2.3. Read section 2.4 to prepare for tomorrow's class.

Wednesday – Class day. After class: rewrite and organize notes, make flash cards. Evening: re-read section 2.4 in the text; Goal: understand examples in text.

Thursday – Study in STEM center. Go over section 2.4 notes. Goal: understand concepts in section 2.4. Begin practice homework for 2.4

Friday – Go over 2.3 and 2.4 notes and flash cards. Goal: Begin memorizing important definitions and formulas. Continue working on section 2.3 and 2.4 practice homework.

Saturday – Go over 2.3 and 2.4 notes and flash cards. Goal: Completely memorize important definitions and formulas. Continue working on section 2.3 and 2.4 practice homework. Meet with study group at 2pm.

Sunday – Make sure all practice homework for Sections 2.3 and 2.4 are complete. Work extra practice problems from the text or online resource. Prepare study plan for next week and read section 2.5 to prepare for tomorrow's class.

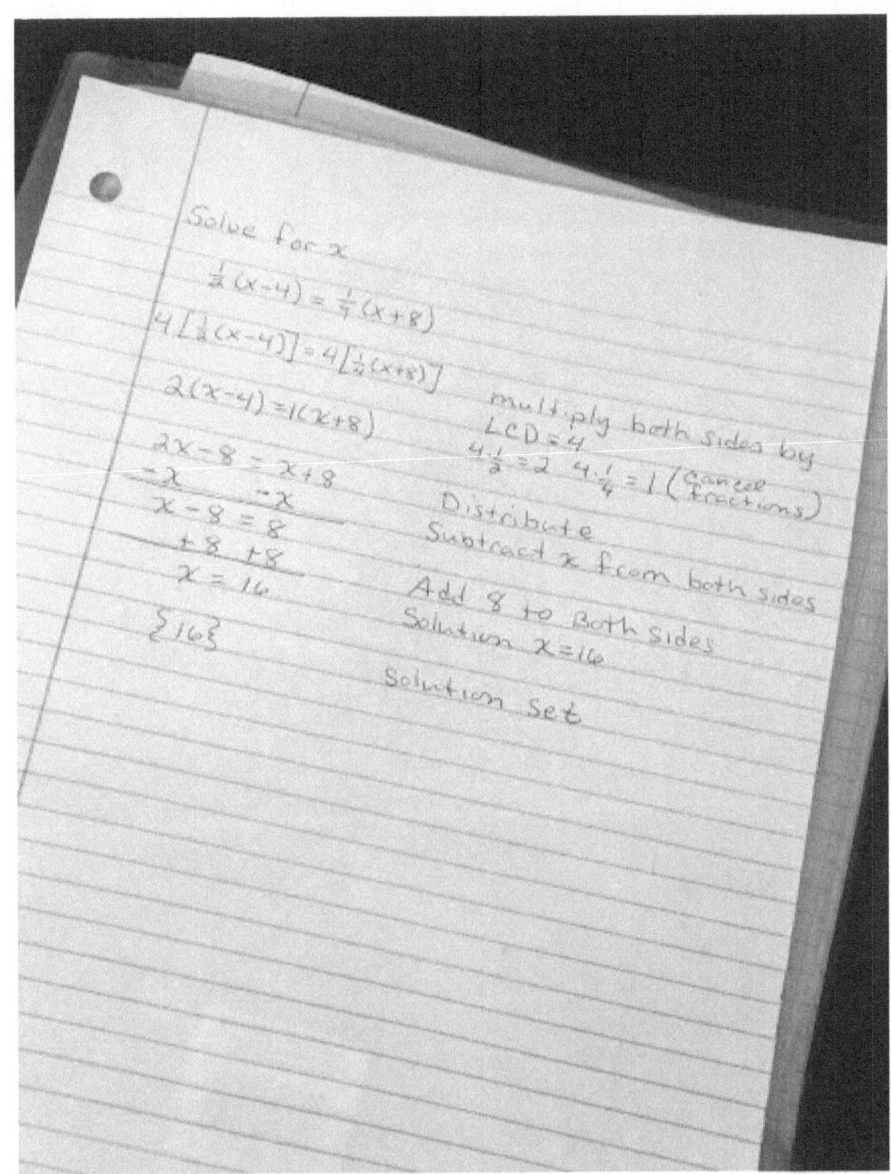

Figure 1. Note-taking Example

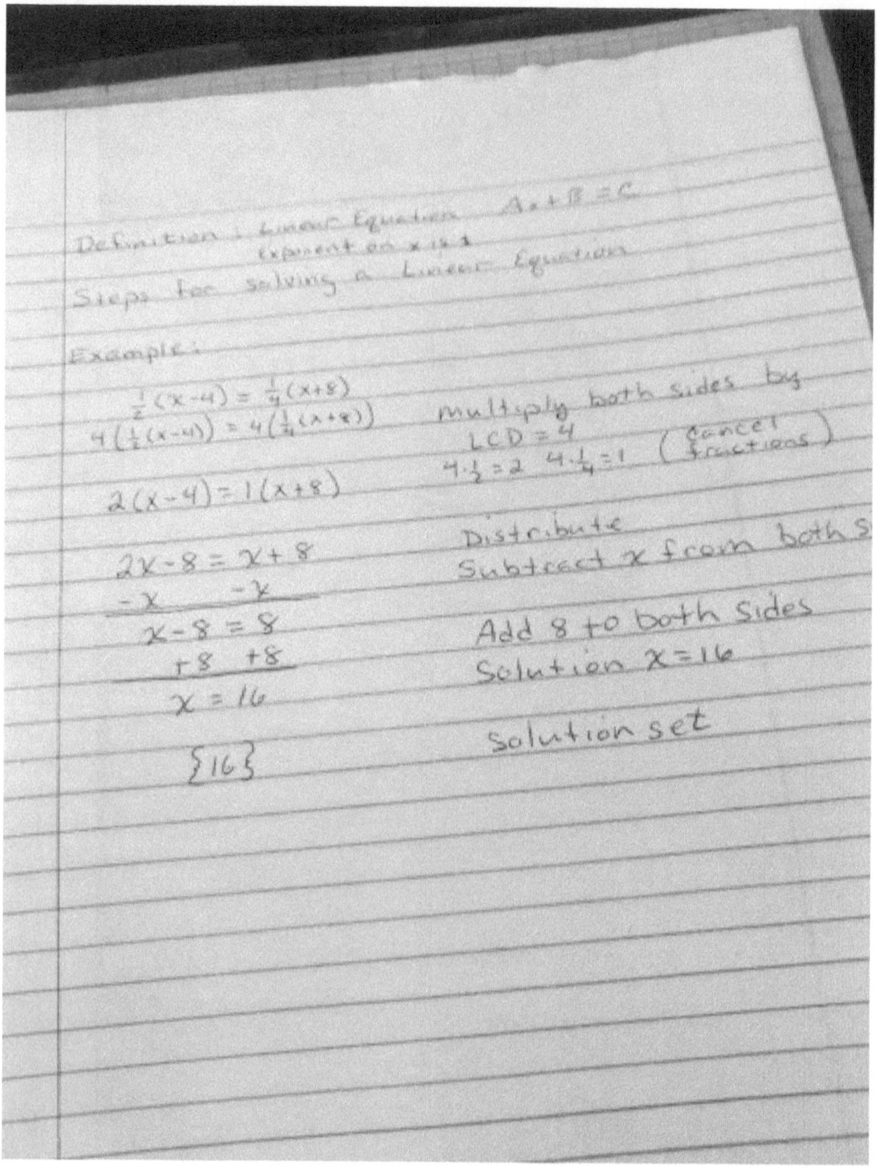

Definition : Linear Equation $Ax + B = C$
         exponent on x is 1
Steps for solving a Linear Equation

Example :

$\frac{1}{2}(x-4) = \frac{1}{4}(x+8)$

$4\left(\frac{1}{2}(x-4)\right) = 4\left(\frac{1}{4}(x+8)\right)$ multiply both sides by
                                         LCD = 4   (cancel
                                   $4 \cdot \frac{1}{2} = 2$ $4 \cdot \frac{1}{4} = 1$ fractions)

$2(x-4) = 1(x+8)$

$2x - 8 = x + 8$ Distribute
$\underline{-x \quad\quad -x}$ Subtract x from both s
$x - 8 = 8$ Add 8 to both sides
$\underline{+8 \quad +8}$ Solution $x = 16$
$x = 16$

$\{16\}$ Solution set

Figure 2a. In-class notes example

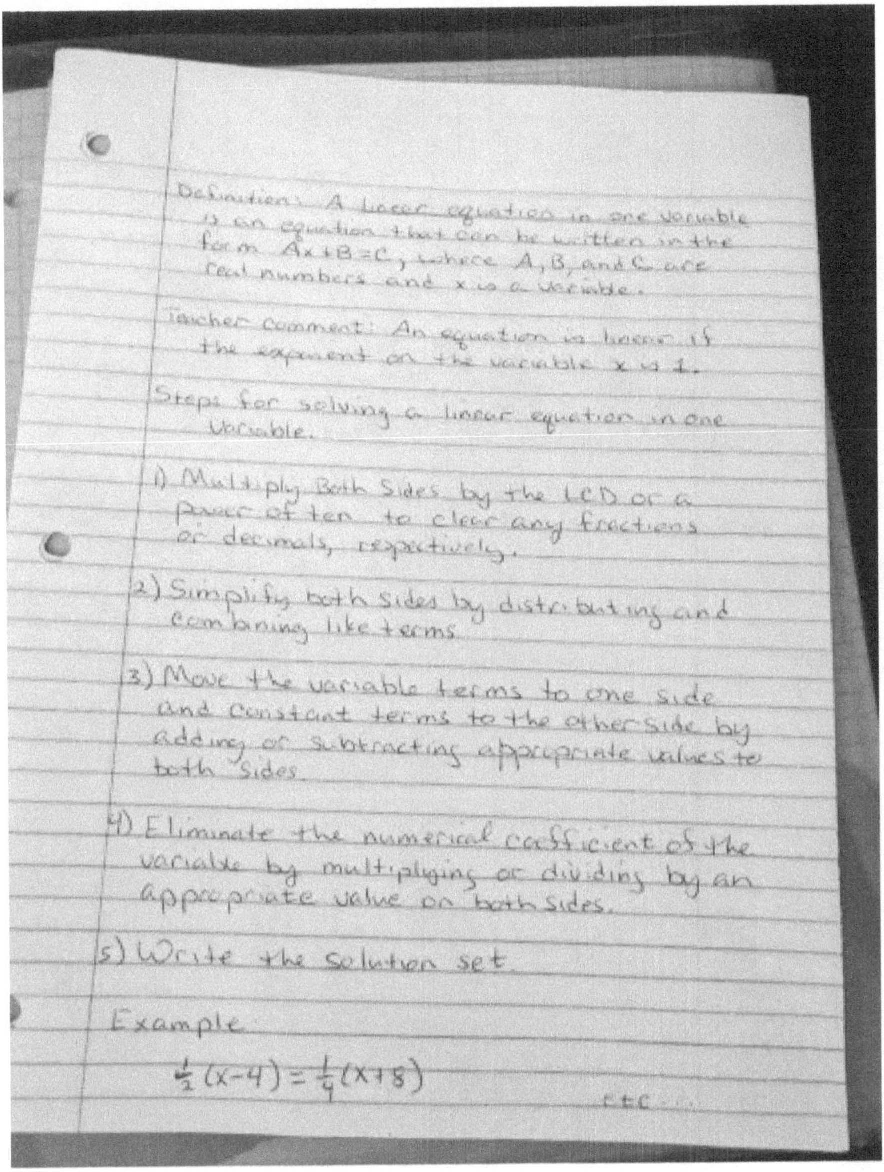

Definition: A linear equation in one variable is an equation that can be written in the form $Ax + B = C$, where A, B, and C are real numbers and x is a variable.

Teacher comment: An equation is linear if the exponent on the variable x is 1.

Steps for solving a linear equation in one variable.

1) Multiply both sides by the LCD or a power of ten to clear any fractions or decimals, respectively.

2) Simplify both sides by distributing and combining like terms.

3) Move the variable terms to one side and constant terms to the other side by adding or subtracting appropriate values to both sides.

4) Eliminate the numerical coefficient of the variable by multiplying or dividing by an appropriate value on both sides.

5) Write the solution set.

Example

$$\frac{1}{2}(x-4) = \frac{1}{4}(x+8)$$

etc...

Figure 2b. After-class notes example

www.ingramcontent.com/pod-product-compliance
Lightning Source LLC
Chambersburg PA
CBHW031559210526
45464CB00003B/1342